I0605337

THE PLANETS OF OUR SOLAR SYSTEM

by Kari A. Cornell

BrightPoint Press

San Diego, CA

an imprint of ReferencePoint Press, Inc.
Printed in the United States

For more information, contact:
BrightPoint Press
PO Box 27779
San Diego, CA 92198
www.BrightPointPress.com

LIBRARY OF CONGRESS CATALOGING-IN-PUBLICATION DATA

Name: Cornell, Kari A., author.
Title: Mars / by Kari A. Cornell.
Description: San Diego, CA: ReferencePoint Press, 2026 | Series: The planets of our solar system | Audience: Grade 7 to 9 | Includes bibliographical references and index.
Identifiers: ISBN: 9781678211646 (hardcover) | ISBN: 9781678211653 (eBook)
The complete Library of Congress record is available at www.loc.gov.

CONTENTS

AT A GLANCE

- Mars is the fourth planet from the Sun.

- Mars is the second-smallest planet in the solar system by diameter. The distance around Mars's equator is around 13,233 miles (21,296 km).

- In 1610, Italian astronomer Galileo Galilei became the first to gaze at Mars through a telescope. He reported that through the lens, the planet looked like a small disk.

- In 1965, the US-launched *Mariner 4* spacecraft became the first to fly past Mars. The probe captured twenty-one photos of a dry and desolate landscape.

- In December 1996, the *Mars Pathfinder* mission launched into space. This mission successfully landed the first Mars rover, *Sojourner*.

- In 1877, American astronomer Asaph Hall discovered the two moons of Mars. He named them Phobos and Deimos.

- The Japan Aerospace Exploration Agency (JAXA) plans for its Martian Moons eXploration (MMX) mission to land on Phobos and collect samples from the moon's surface.

LANDING IN THE CRATER

The moment had finally come. It was February 18, 2021. Engineers and scientists gathered at the Jet Propulsion Laboratory (JPL). This laboratory is in Pasadena, California. It is run by the National Aeronautics and Space Administration (NASA).

This team of people had created the *Perseverance* rover. A rocket had carried

The *Perseverance* rover is about 10 feet (3 m) long, making it the size of a small car.

Members of the *Perseverance* team at JPL celebrate the rover's successful landing on Mars.

the rover into space months ago. Now the rover was about to land on Mars.

Everyone stared at the screens in front of them. The spacecraft had cameras on it. The team could see the rover from multiple angles. Soon an orange-and-white parachute appeared. A woman talked over a loudspeaker. She said, “We are seeing significant deceleration in the velocity.” This meant the parachute was working. It was slowing down the rover so that it could make a safe landing on Mars.

The rover had a built-in navigation system. The system took pictures as the rover approached Mars. It compared these photos to maps of the Martian surface. This helped guide the rover to a safe landing spot.

Ridges and craters came into view. Soon the force of retro rockets blew red dust across the landing site. These rockets quickly slowed the rover’s descent. Team members at JPL jumped up and began to cheer. The rover had landed safely on Mars.

EXPLORING MARS

Perseverance's mission was to explore Mars's Jezero Crater. It searched for signs of ancient life. The rover found igneous rocks. These rocks form when **magma** or lava cools. Scientists study igneous rocks to learn about the environment in which they formed. They hope to study samples of Martian igneous rocks. This could tell them more about the history of the Jezero Crater.

The rover's cameras took hundreds of thousands of photos. *Perseverance* also had a small helicopter on board named *Ingenuity*. The helicopter had two cameras. It made seventy-two flights.

Perseverance and *Ingenuity* are not the first spacecraft to explore Mars. Many others have orbited and landed

***Ingenuity* flew as high as 78.7 feet (24 m) during one of its flights.**

on the planet. They've taken pictures and gathered data. This information has helped scientists learn about Mars. They've learned about its landscape, atmosphere, and more.

CHAPTER ONE

MEET THE RED PLANET

People have been studying Mars for thousands of years. Today, rovers such as *Perseverance* help scientists learn about the planet. In the past, people watched and wondered about Mars from Earth.

Mars is the fourth planet from the Sun. It is the second smallest planet in the solar system by diameter. The distance around the planet's equator is around 13,233 miles (21,296 km). Mars has about 10 percent

People have taken countless pictures of Mars. This picture was taken in 2003 by a spacecraft called the *Mars Global Surveyor*.

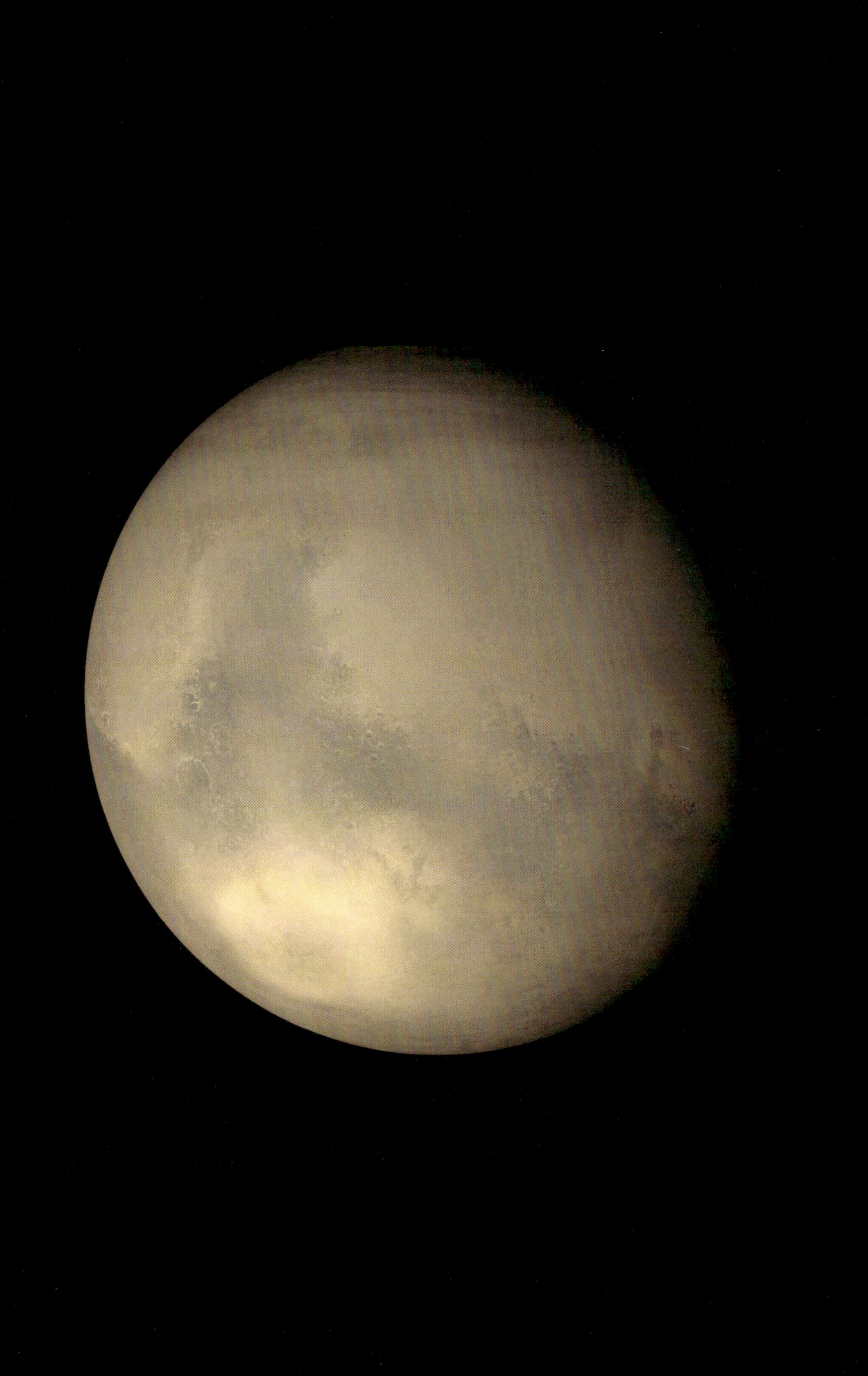

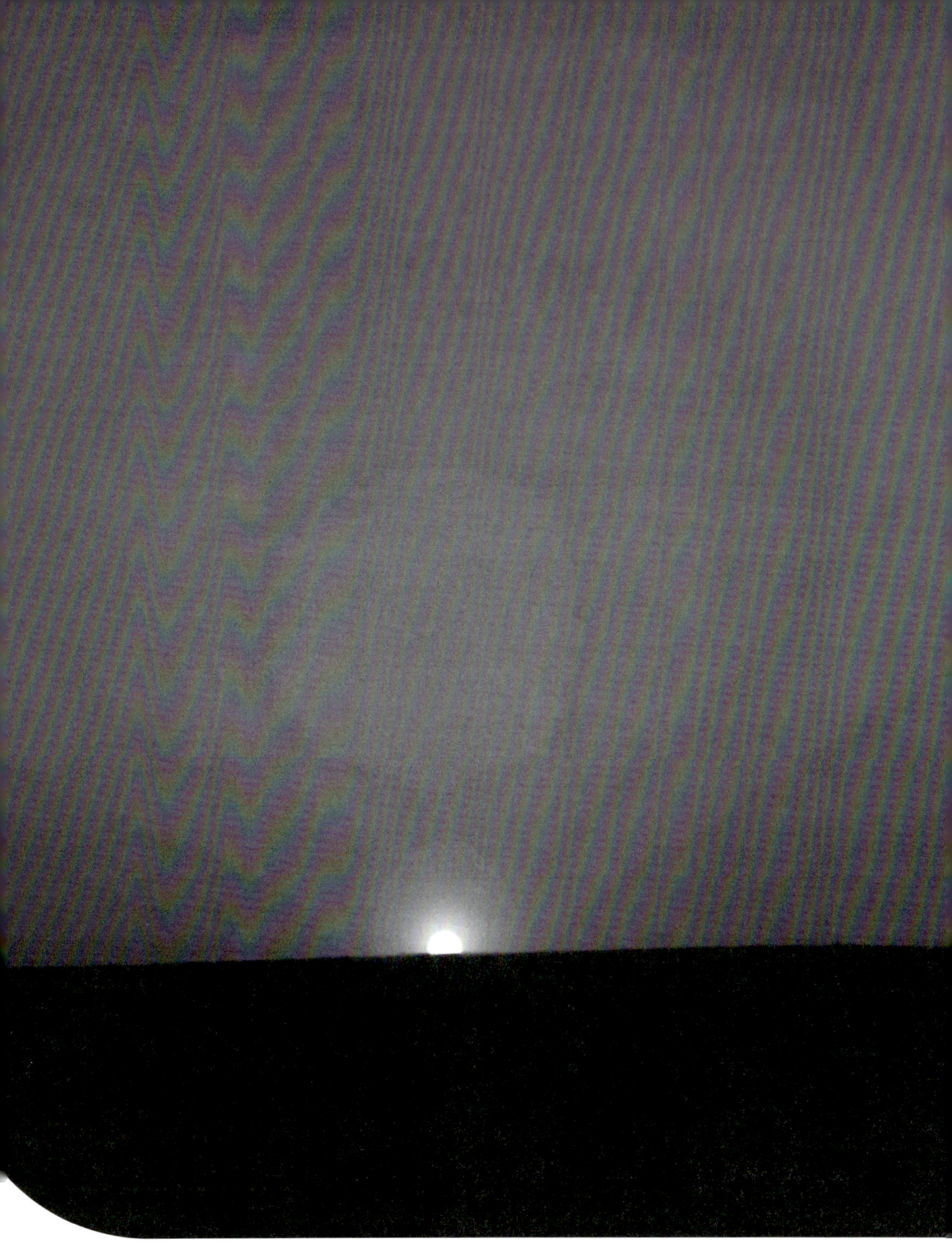

NASA's *Phoenix* lander took this picture of the Sun rising over Mars in 2008.

the mass of Earth. But it has about 15 percent the volume of Earth. That means Mars is less dense than Earth. Density is a measure of how tightly packed something is. Scientists think this difference comes from lighter material in Mars's core.

Mars does not orbit the Sun in a perfect circle. The planet is closer to the Sun at certain times. The planet is located an average of 142 million miles (229 million km) from the Sun. It takes sunlight 13 minutes to reach Mars at this distance. Mars's orbit lies between those of Earth and Jupiter.

MARS AND EARTH

Today, the surfaces of Mars and Earth are quite different. Mars is a rocky, cold desert. The planet has a very thin atmosphere. It is mostly made up of the gas carbon dioxide. Temperatures on Mars range between about –225°F and 70°F (–143°C and 21°C). People from Earth could not survive there without life-supporting technology.

Another reason people would struggle to live on Mars is its lack of a full **magnetic** field. Magnetic fields are regions of magnetic forces. They come from movement of material beneath some

How Cold Is It?

The thin atmosphere on Mars means heat from the Sun escapes quickly. People standing on Mars's equator at noon would be able to feel this. Their feet would feel warm, as they might on a spring day. But their heads would feel as if they were in an icebox.

planets' surfaces. These planets include Earth. Earth's magnetic field shields the planet from dangerous space material. Some of this material comes from the Sun. This material can strip away a

The Sun constantly ejects material called solar wind. Earth's magnetic field protects the planet from the worst effects of this material.

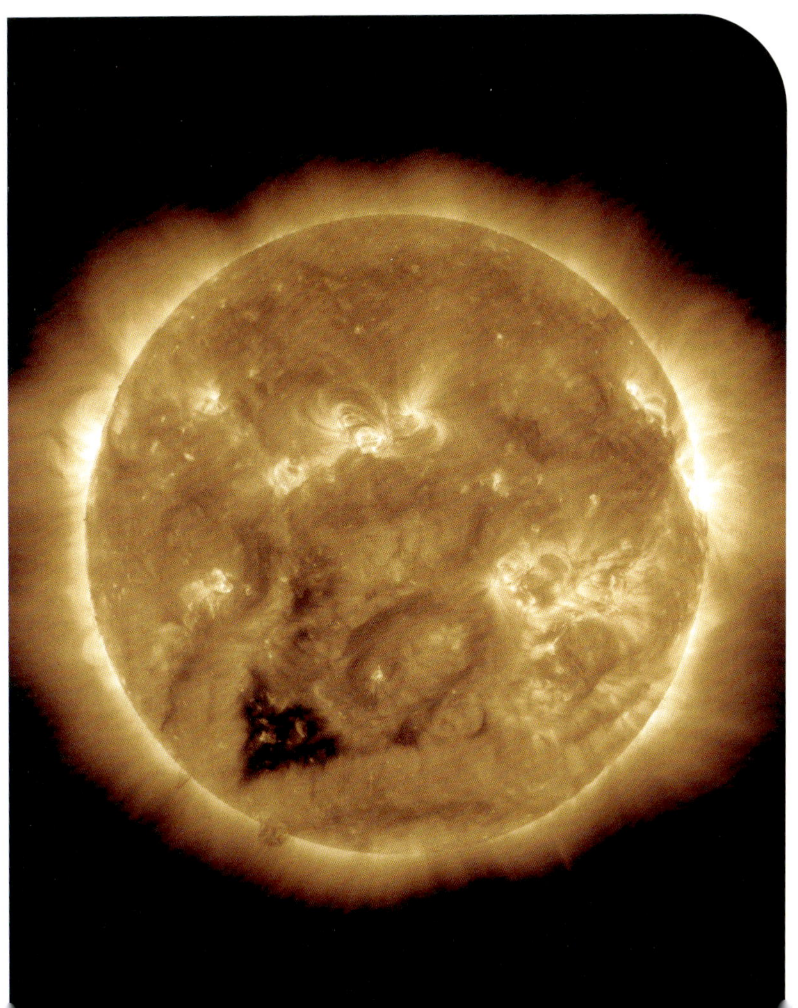

As on Earth, dust devils occasionally rip across Mars's surface. Dust devils are small whirlwinds that kick up dust.

planet's atmosphere. Life as it is today would not exist without this field.

Mars's desert surface is prone to windstorms. These huge storms create clouds of dust. They can last for weeks. Rare gigantic storms can cover the whole planet. Michael Smith is a planetary scientist for NASA. He says, “Once every three Mars years (about 5.5 Earth years), on average, normal storms grow into planet-encircling dust storms.”[1]

But evidence suggests that Mars was not always this way. One of the reasons scientists study Mars is that it has some similarities to Earth. Evidence suggests Mars once had bodies of water, too. NASA has been researching Mars for more than 60 years. Scientists have discovered signs that Mars had rivers and lakes billions of years ago. Mars had a much thicker atmosphere then. This made the planet warmer. Conditions may have been able to support life.

Mars has an **axis of rotation** of 25 degrees. This is a little more than Earth’s tilt of 23.4 degrees. It takes a similar amount of time for Mars and Earth to complete one rotation on their axis. It takes Mars 24 hours and 37 minutes to rotate once. This is the length of one day on Mars. Earth completes one rotation in about 24 hours.

The degree of tilt affects the planet's seasons. Mars and Earth both have four seasons. But the seasons on Mars last longer than Earth's seasons. Martian seasons also vary in length between the northern and southern **hemispheres**. Southern summers are shorter than those in the north.

A TERRESTRIAL PLANET

Like Earth, Mars is a terrestrial planet. Terrestrial planets share common features.

The dunes near Mars's north pole are covered by seasonal caps of dry ice. In the summer, most of this ice disappears.

One is that they have a hard surface made up of rocks or metals. These planets all have a core, mantle, and crust. The core is the hot metal center. The mantle is a layer of solid rocks, minerals, and magma. The crust is the hard outer surface. Mars has a core made of iron, nickel, and sulfur. Beyond the core is a thick mantle. Mars's crust is made up of iron, magnesium, and other elements.

Terrestrial planets also share common landforms. These include volcanoes and valleys. They also include craters and canyons. On Mars many of these landforms are made up of red rocks. But some rocks are brown and gold. Mars's Olympus Mons is the solar system's biggest volcano. It is three times the height of Earth's Mount Everest.

Mars also has vast canyons. Valles Marineris is about 2,500 miles (4,000 km) long. The canyon system is up to

375 miles (604 km) wide at one point. This part of the canyon plunges about 5.6 miles (9 km) into the Martian crust. The canyon is known as the Grand Canyon of Mars. It is more than nine times the length of Earth's Grand Canyon.

Valles Marineris is so large that it is visible from space.

CHAPTER TWO

MARS THROUGH A TELESCOPE

The first people to study Mars did so from Earth. Mars is easy to see with the naked eye. It looks a little larger than a star. It also has a slight pink or red tint. And the planet moves in the sky differently than stars do.

People in the early 1600s began to use telescopes to study the night sky. Galileo Galilei was an Italian astronomer. In 1610, he became the first to gaze at Mars through

Galileo Galilei was born on February 15, 1564, in Pisa, Italy.

GALILEO GALILEI

Mars is closer to Earth at certain times. On nights when Mars is especially close, the planet appears much brighter in the sky than normal.

a telescope. He reported that the planet looked like a small disk.

One of the first things skywatchers noticed about Mars was its red color. This is why Mars is sometimes called the Red Planet. Mars's color inspired its name. Mars is the Roman god of war. The planet's red color reminded people of blood.

Mars is red because the dust and rocks that make up its surface contain iron.

Iron rusts when it is exposed to oxygen. This gives Mars its color. Penny Wozniakiewicz is a planetary scientist. She says, "[Mars's] atmosphere also exhibits a reddish hue."[2] She explains that this is because some dust from Mars's surface floats in its atmosphere.

The earliest astronomers were puzzled by Mars's movement. Stars appeared to move in consistent arcs. But planets appeared to move irregularly. Mars sometimes followed the Sun. Other times it moved in the opposite direction.

People long believed Earth was the center of the universe. They thought the Sun and planets revolved around Earth. Astronomer Nicolaus Copernicus suggested a new idea in 1515. He said that Earth and the planets all revolve around the Sun.

Copernicus published the final version of his theory just before he died in 1543. The idea helped explain the odd way Mars moved through the sky. The planet was orbiting the Sun, not Earth.

THE MARTIAN SURFACE

Through the lens of a telescope, much of Mars looks bright reddish-orange. Early observers thought these areas were deserts. There were also dark patches around the north and south poles. Another large dark patch could be seen in the center of the planet. Astronomers could see this patch at certain times of the year. These dark patches were thought to be oceans.

In 1659, Dutch scientist Christiaan Huygens studied the Red Planet's surface.

This statue in Poland depicts Copernicus holding his model of the solar system with the Sun in the middle.

He had improved the lens of his telescope. This helped him see the darkened area in more detail. The area came to be known as Syrtis Major. At first this feature was thought to be a shallow sea. Scientists later learned that the area was actually a type of volcano. It measured 620 miles (1,000 km) across. And it was 930 miles (1,500 km) from top to bottom. Huygens then made the first

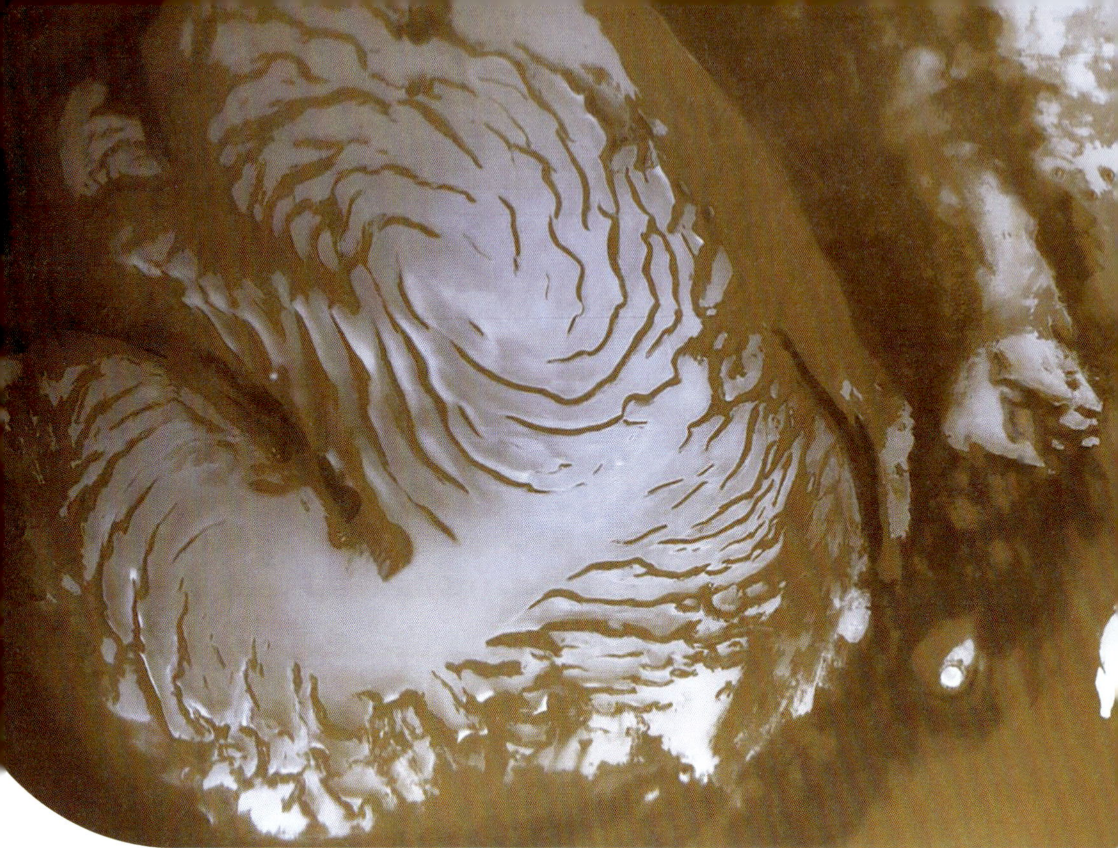

Mars's north polar cap remains year-round, surviving the Martian summer.

drawings of the surface of Mars. This rough map included Syrtis Major.

Italian-French astronomer Gian Domenico Cassini made another discovery in 1666. He noted that the dark patches around the poles were polar caps. William Herschel continued to study these features in the 1780s. He was a British astronomer who discovered Uranus in 1781. His telescopes allowed him to see the caps in greater detail.

In 1783, Herschel made drawings of the cap on the South Pole. He drew the cap on different dates over a year. Its size changed over time. This meant Mars had different seasons. Herschel also measured the tilt of Mars's rotation on its axis.

The Martian Polar Caps

William Herschel believed the Martian polar caps were made of water ice, just as on Earth. But in 1898, George J. Stoney suggested the polar caps were carbon dioxide. Astronomer Gerard Kuiper detected carbon dioxide in Mars's atmosphere in 1947. Later scientists showed that the polar caps contain both water ice and carbon dioxide.

MAPPING MARS AND BEYOND

Herschel's drawings were the first maps made of the polar caps. German scientists Wilhelm Beer and Johann Heinrich von Mädler created the first map of the entire planet in 1830. The map showed the light and dark areas of Mars.

Giovanni Schiaparelli of Italy created the first modern map of Mars in 1877. Schiarparelli believed he saw *canali* on Mars's surface. This Italian word means *channels* in English. But some sources translated *canali* as *canals*. People took this to mean that there may have been intelligent life on Mars that built these canals. Observers later discovered that the channels were an illusion.

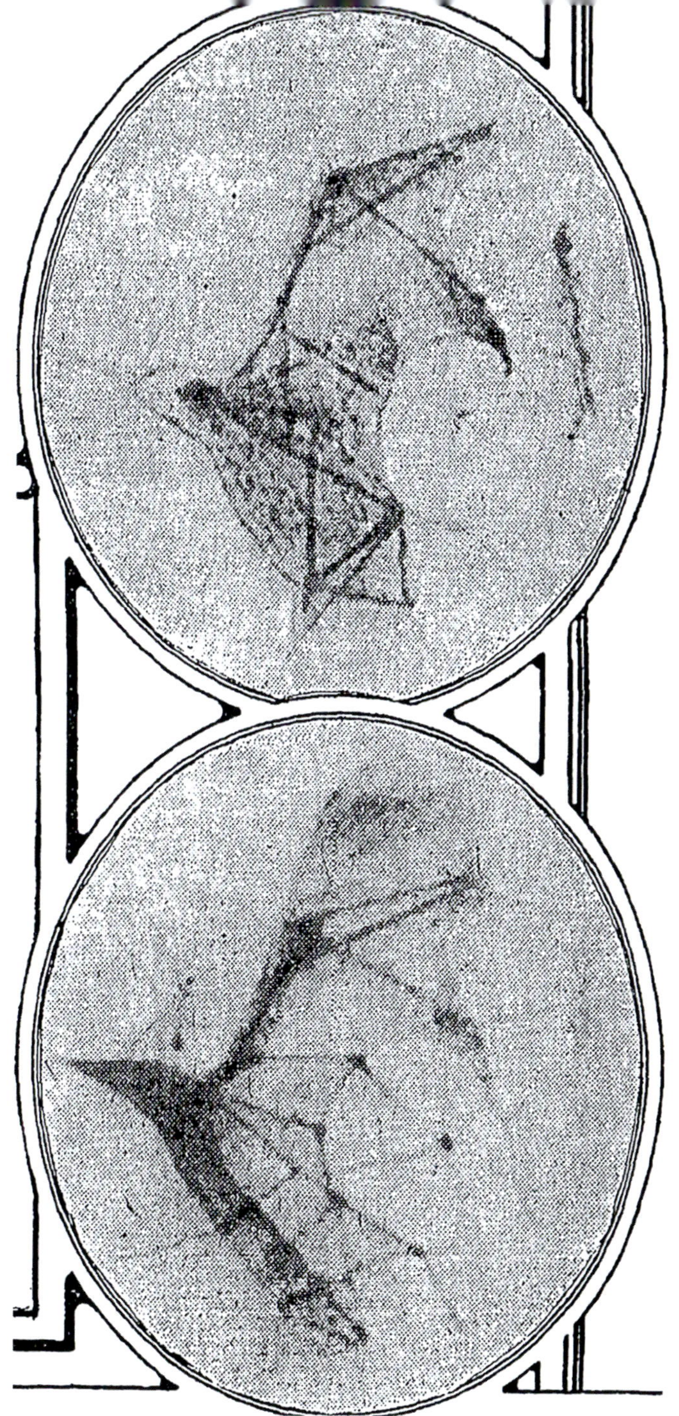

American astronomer Percival Lowell produced this drawing of Martian canals that he believed had been created by intelligent life.

CHAPTER THREE

EXPLORING MARS

Astronomers still observe Mars from Earth. But now there are many other ways to learn about the Red Planet. Spacecraft orbit Mars to take photos. Rovers such as *Perseverance* study the planet's surface. Mars is the most explored planet in the solar system besides Earth.

The exploration of Mars began in the 1960s. The United States launched

An Atlas V rocket carried the *Perseverance* rover into space on its trip to Mars.

***Mariner* 4 made its closest approach to Mars on July 15, 1965.**

Mariner 4 in 1965. It became the first spacecraft to fly past Mars. It captured twenty-one photos of a dry and empty landscape. *Mariners 6* and *7* also completed successful flybys. The two probes were launched in 1969. *Mariner 9* orbited Mars and captured photos of a dust storm. It circled the planet from November 1971 to October 1972.

Mariners 6, *7*, and *9* provided scientists with a lot of information. They learned that the Martian surface was shaped by volcanoes, internal forces, and water. They also discovered Valles Marineris. These spacecraft gathered data about Mars's atmosphere as well.

TOUCHING DOWN

Landing a spacecraft on Mars is hard. Nadia Drake is a science journalist. She explains, "The thin Martian atmosphere makes descent tricky, and more than 60 percent of landing attempts have failed."[3] The Soviet Union managed to land the first operational spacecraft on Mars.

MISSIONS TO MARS

1964: ***Mariner 4.*** This probe captured the first pictures of Mars from deep space.

1969: ***Mariner 6* and *Mariner 7*.** These spacecraft took additional pictures of Mars.

1971: ***Mariner 9.*** This spacecraft was the first to orbit a planet besides Earth.

1975: ***Viking* Project.** The *Viking* Project resulted in the first safe US landing on Mars.

1996: ***Mars Pathfinder.*** This spacecraft successfully landed the *Sojourner* rover on Mars.

2001: ***Mars Odyssey.*** This spacecraft orbited Mars and produced a map of the minerals that make up its surface.

2003: **Mars Exploration Rover.** This mission brought the *Spirit* and *Opportunity* rovers to Mars.

2005: ***Mars Reconnaissance Orbiter.*** This orbiter investigated the history of water on Mars.

2011: **Mars Science Laboratory.** This mission landed the *Curiosity* rover on Mars.

2013: ***MAVEN.*** The *MAVEN* orbiter studied Mars's atmosphere from space.

2018: ***InSight.*** The *InSight* lander studied Mars's interior structure.

2020: **Mars 2020.** This mission brought the *Perseverance* rover and the *Ingenuity* helicopter to Mars.

This timeline lists NASA's major missions to Mars between 1964 and 2020.

On December 2, 1971, *Mars 3* landed on the Red Planet. It touched down during a dust storm. *Mars 3* managed to capture about 20 seconds of data. Then, the transfer of data suddenly stopped. The dust storm might have caused the lander to fail.

Two US spacecraft reached Mars in 1976. They were *Viking 1* and *Viking 2*. Each had an orbiter and a lander. The twin landers successfully touched down on Mars. The missions focused on looking for signs of life. But they found no evidence of it. They did discover that the makeup of the Martian surface matched some **meteorites** that had been found on Earth. Scientists believe those meteorites may have come from Mars.

ROBOTIC ROVERS

In December 1996, NASA launched the *Mars Pathfinder* mission. The goal was to land the first Mars rover. It was called *Sojourner*. NASA used a new landing method this time. First, a parachute slowed its descent. Airbags around the rover inflated as it neared the surface. The rover bounced 15 times on the Martian surface. One bounce was as high as 50 feet (15.2 m). But it eventually slowed to a stop. The airbag panels opened to reveal *Sojourner* inside. The rover rolled off the lander. It was soon communicating with scientists on Earth.

The team that worked on the rover was excited. Brian Muirhead was the flight systems manager. He said, "We are on the

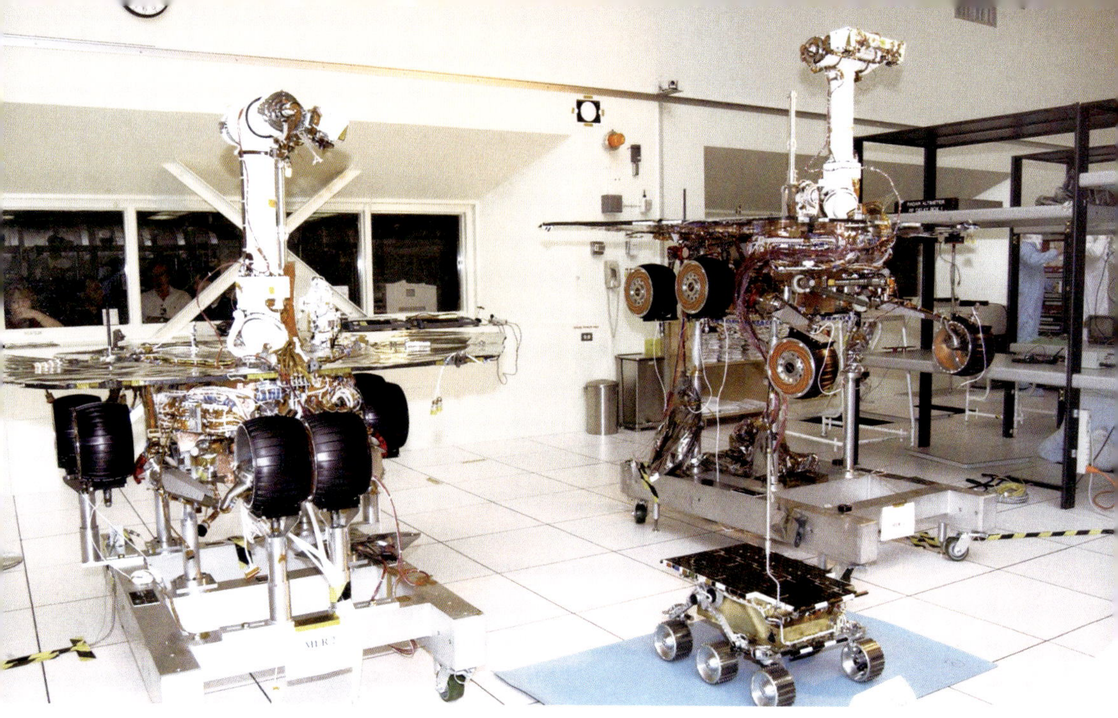

***Sojourner* (middle) stands between *Spirit* (left) and *Opportunity* (right).**

surface of Mars . . . This is way beyond our expectations."[4] During the mission, the lander captured 16,500 images. The rover took 550 photos. It also gathered rock and soil samples. It collected data on the weather. NASA scientists learned a lot about Mars from this mission. At one time the planet was warmer and wet.

Other rovers followed *Sojourner*. The *Spirit* and *Opportunity* rovers landed on

Mars in January 2004. They explored opposite sides of the planet. *Curiosity* launched in 2012. It studied Gale Crater. This was the site of an ancient lake. The *InSight* lander began exploring a vast plain

The Search for Signs of Ancient Life

The *Perseverance* rover discovered an interesting rock in July 2024. The rock has white spots. Astrobiologist David Flannery says, "These spots are a big surprise. On Earth, these types of features in rocks are often associated with the fossilized record of **microbes** living in the subsurface." *Perseverance* took a core sample of the rock. These samples will be returned to Earth on a future mission.

Quoted in "NASA's Perseverance Rover Scientists Find Intriguing Mars Rock," NASA, *July 25, 2024. www.nasa.gov.*

along Mars's equator in November 2018. The *InSight* lander studied volcanic activity on Mars. It detected quakes. It also found liquid water deep underground.

EXPLORATION CONTINUES

Many spacecraft continue to orbit Mars and send back photos. Several are from NASA. These include *MAVEN* and the *Mars Reconnaissance Orbiter*. The European Space Agency (ESA) has two missions. They are *Mars Express* and the *ExoMars Trace Gas Orbiter*. China has the *Tianwen-1* orbiter and the *Zhurong* rover. They searched for water beneath the surface of Mars. *Hope* was a Mars mission launched by the United Arab Emirates. It studied the Martian climate.

People living on Mars would have to deal with a lack of oxygen, dangerous sun rays, extreme temperatures, and other factors.

NASA plans to send humans to Mars by the 2030s. It is working on a space capsule called *Orion* for the mission. Private company SpaceX has a plan to build a city on Mars by the 2040s. The company is designing spacecraft and technologies to support human life on Mars.

However, others doubt SpaceX's plan. Sending people to Mars would be hard. It would be very expensive. And the trip to Mars would take months. People worry about danger to the crew and to Mars itself. Scientists are working to come up with solutions.

THE MOONS OF MARS

Asaph Hall made an important discovery in 1877. Hall was an astronomer with the US Naval Observatory. He noticed that Mars appeared to have two natural **satellites**. Hall named the moons Phobos and Deimos. These were the sons of Ares, the Greek god of war.

Little was learned about Mars's moons for almost 100 years. But that changed in 1971. *Mariner 9* took photos of the moons.

Asaph Hall uses the telescope with which he discovered Mars's moons.

The crater on Phobos that *Mariner 9* took pictures of is called Stickney Crater.

These photos revealed a crater on Phobos that measured 6 miles (10 km) wide. The images taught scientists much more about the moons.

Both moons are shaped more like potatoes than spheres. Neither has enough mass for gravity to make it round. They are also believed to be among the smallest moons in the solar system. The moons are mostly made of carbon-based rocks and ice. They are pocked with several craters. The moons are also covered with dust and loose rocks. Neither moon reflects very much light.

PHOBOS

Phobos is the larger of the two Martian moons. It measures 17 miles (27 km) across on its longest side. It is closer to Mars than Deimos. It orbits about 3,700 miles (6,000 km) above the planet on average. The larger moon orbits the planet once every 7 hours and 39 minutes. That's three times each Martian day.

Phobos is moving toward Mars very slowly. It gets about 5.9 feet (1.8 m) closer

***Perseverance* took this picture of Phobos eclipsing the Sun from Mars.**

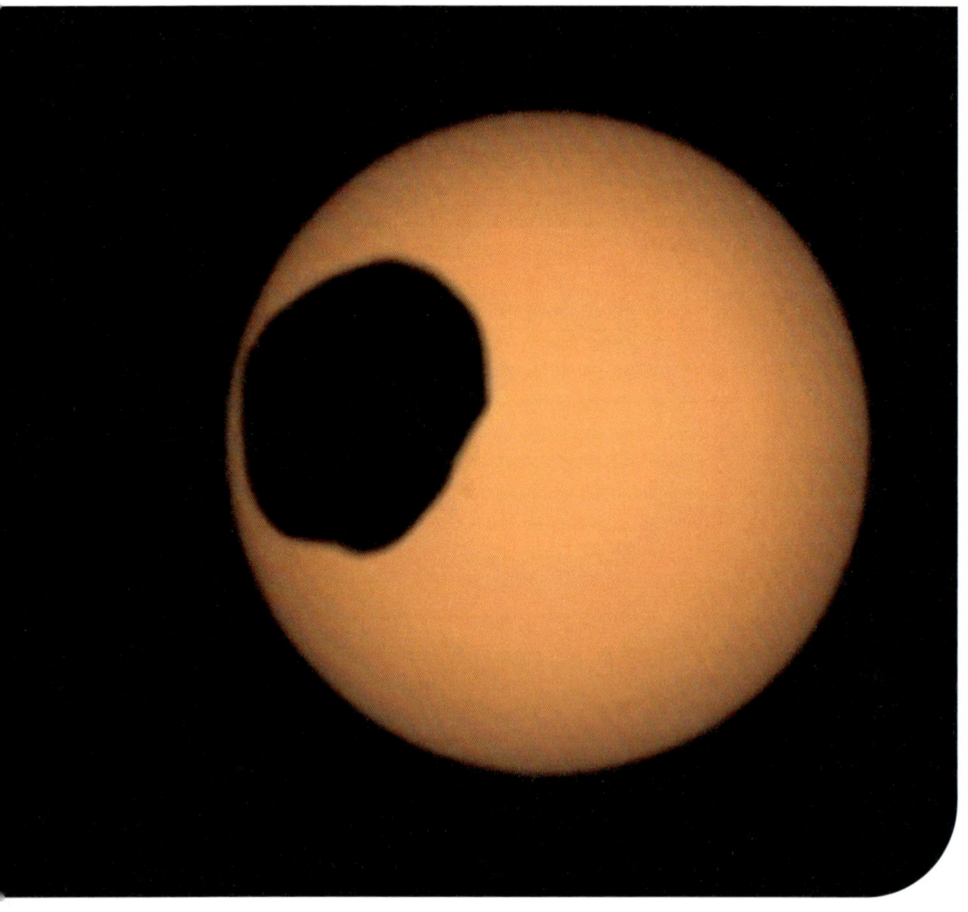

to Mars every 100 years. The moon will probably crash into the planet within 50 million years. It could also break into pieces and form a ring around Mars.

DEIMOS

Mars's smaller moon is Deimos. It measures about 9 miles (15 km) across on its longest side. It orbits about 14,500 miles (23,500 km) from Mars on average. The moon takes 30 hours and 17 minutes to circle Mars. Unlike Phobos, Deimos is pulling farther away from Mars. Scientists predict that at some point the moon will fly off into space.

Deimos is light gray and has many craters. The largest two craters measure about 1.6 miles (2.5 km) across. They are named Swift and Voltaire. The craters were formed by meteors. But the moon is relatively smooth compared to Phobos. This is because a fine layer of rocky **debris** covers the surface of the smaller moon. This debris comes from meteorites that break apart as they hit the moon. The rocky layer is as deep as 328 feet (100 m) in some places.

EXPLORATION

A few missions have set out to gather data on Mars's moons. The first were two Soviet spacecraft called *Phobos 1* and *2*. They were launched in 1988. *Phobos 1* failed on its way to Mars. But *Phobos 2* arrived in 1989. It took photos and gathered data during flybys of the two moons.

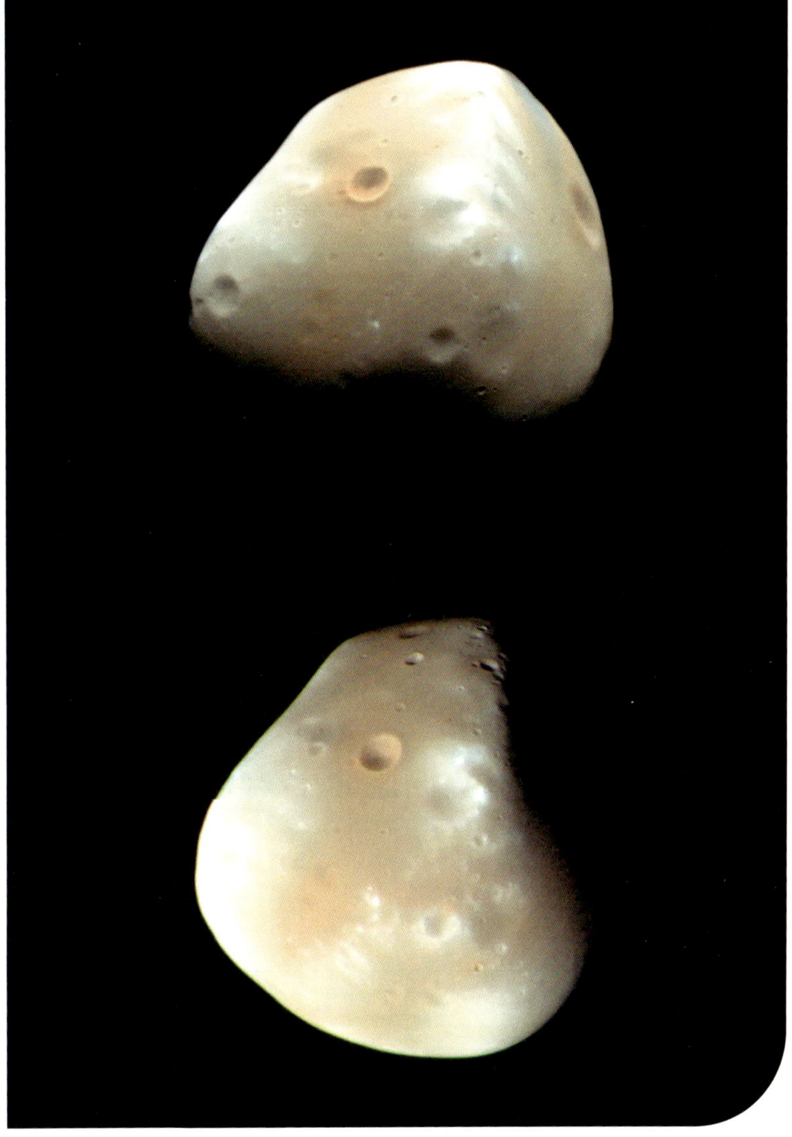

Deimos is tidally locked to Mars. That means the same side of Deimos always faces Mars.

A Russian mission launched on November 8, 2011. It was called *Fobos-Grunt*. Its mission was to bring soil samples from Phobos back to Earth.

Asteroids are space rocks that orbit the Sun.

But the rockets designed to lift the craft out of Earth's orbit failed. It fell into the Pacific Ocean on January 15, 2012.

More recent missions have provided lots of data. The two moons of Mars have long been thought to be captured asteroids. But evidence gathered by the *Hope* orbiter tells a different story. *Hope* has been orbiting Mars since 2021. The orbiter provided photos of the side of Deimos that faces away from Mars in 2023. This side had never been seen before. The photos suggest Deimos is made of material that is similar to Mars. This means the moons may have formed from part of the planet itself.

Scientists have come up with several ways this could have happened. A planetary body may have struck Mars and created the moons. Or the moons may be parts of an earlier moon that somehow broke apart. “This is one of the big, outstanding mysteries in planetary science,” says Abigail Fraeman, a NASA planetary scientist. “However [Phobos and Deimos] formed, there are broad implications for our understanding of the solar system.”[5]

The Japan Aerospace Exploration Agency (JAXA) hopes to solve the mystery. Its Martian Moons eXploration (MMX) mission will launch a solar-powered spacecraft in 2026. It will enter Mars's orbit in 2027 if all goes as planned. Its mission will be to land a rover on Phobos and collect samples from its surface.

Sticking the Landing

Keeping a rover on Phobos is challenging. Phobos has one-thousandth the gravity of Earth. "If we're going quicker than 80 millimeters per second [(3.1 inches per second)], we might flip over the rover or even leave the Phobos system," explains Dr. Markus Grebenstein.

Quoted in Jonathan O'Callaghan, "Where Did Mars's Moons Come From?" Scientific American, *May 2, 2023. www.scientificamerican.com.*

NASA and JAXA scientists work on an instrument that will go on the MMX spacecraft.

The goal is to return samples to Earth for study by 2031. As scientist Jemma Davidson says, "Understanding how the moons formed is really fundamental to us understanding the dynamics of our solar system."[6]

GLOSSARY

axis of rotation

an imaginary line running through a planet's poles around which the planet rotates

debris

loose fragments of rock

hemispheres

halves of a sphere

magma

molten rock beneath a planet's surface

magnetic

related to the motion of electric charges

meteorites

space rocks that strike a planet

microbes

very small living things that cause disease

satellites

objects that orbit other objects of larger size

SOURCE NOTES

CHAPTER ONE: MEET THE RED PLANET

1. Quoted in Karl B. Hille, "The Fact and Fiction of Martian Dust Storms," *NASA*, September 18, 2015. www.nasa.gov.

CHAPTER TWO: MARS THROUGH A TELESCOPE

2. Penny Wozniakiewicz, "Why Is Mars Red?" *BBC Sky at Night Magazine*, November 16, 2023. www.skyatnightmagazine.com.

CHAPTER THREE: EXPLORING MARS

3. Nadia Drake, "Why We Explore Mars—and What Decades of Missions Have Revealed," *National Geographic*, n.d. www.nationalgeographic.com.

4. Quoted in *CBS Evening News*, "From the Archives: NASA's Mars Pathfinder Space Probe Lands on Mars," *YouTube*, July 4, 2023. www.youtube.com.

CHAPTER FOUR: THE MOONS OF MARS

5. Quoted in Tom Metcalfe, "A New Look at Mars' Moon Deimos Highlights Its Mysterious Origin," *Astronomy*, May 18, 2023. www.astronomy.com.

6. Quoted in Jonathan O'Callaghan, "Where Did Mars's Moons Come From?" *Scientific American*, May 2, 2023. www.scientificamerican.com.

FOR FURTHER RESEARCH

BOOKS

Dr. Bruce Betts, *Solar System Reference for Teens*. Callisto, 2022.

Tammy Gagne, *Colonizing Mars*. BrightPoint Press, 2023.

Meg Marquardt, *Earth*. BrightPoint Press, 2026.

INTERNET SOURCES

Jason Davis, "Mars, the Red Planet," *The Planetary Society*, May 2024. www.planetary.org.

"Humans in Space," *NASA*, July 16, 2024. www.nasa.gov.

"October 4, 1957 CE: USSR Launches Sputnik," *National Geographic*, October 19, 2023. https://education.nationalgeographic.org.

WEBSITES

NASA: Mars
https://science.nasa.gov/mars

This NASA website gives visitors an overview of Mars. It answers questions such as how NASA has explored Mars and how it plans to explore it in the future.

NASA: Mars 2020: Perseverance Rover
https://science.nasa.gov/mission/mars-2020-perseverance

The Mars 2020: Perseverance Rover website from NASA explains the *Perseverance* mission. It gives information about what scientists have learned from *Perseverance*'s activities on Mars.

The Schools' Observatory
www.schoolsobservatory.org

The Schools' Observatory provides educational resources related to space. It also informs visitors about space career paths.

INDEX

IMAGE CREDITS

Cover: © NASA
5: © NASA
7: © NASA
8: © Bill Ingalls/NASA
11: © NASA
13: © NASA
14: © NASA
17: © NASA
18: © NASA
21: © NASA
23: © NASA
25: © Anamaria Mejia/Shutterstock Images
26: © Kim Shiflett/NASA
29: © Tomasz Bidermann/Shutterstock Images
30: © NASA
33: © Percival Lowell and E. C. Slipher/The New York Times
35: © Joel Kowsky/NASA
36: © NASA
38: © Red Line Editorial
41: © NASA
44: © NASA
47: © National Photo Company Collection/Library of Congress
48: © NASA
50: © NASA
53: © NASA
54: © NASA
57: © Ed Whitman/NASA

ABOUT THE AUTHOR

Kari A. Cornell is an award-winning author who gardens, runs, and makes pottery. She lives in Minneapolis with her husband and their sweet dog, EmmyLou.